FRONT SIGHT FOCUS

Ten phrases US Navy SEALs use to ensure mission success.

*Follow your Passion.
Greatness Awaits!*

David Havens

FRONT SIGHT FOCUS

Ten phrases U.S. Navy SEALs use to ensure mission success.

David Havens

ETA Solutions LLC

Educate · Train · Anticipate

ETA Solutions

Educate · Train · Anticipate

Educating and *Training* today's leaders to *Anticipate* the next move.

www.ETASolutionsLLC.com

Copyright © 2016 by David Havens

All rights reserved. No part of this publication may be reproduced, stored in a retrieval system, or transmitted, in any form or by any means, electronic, mechanical, photocopying, recording, or otherwise, without prior written permission from the author.

ISBN: 978-0-9982959-0-9

Book cover design by David Havens and Daniel Nestor

Editors: Daniel and Shelby Nestor

Printed in the United States of America

To my beautiful wife and three incredibly beautiful, smart and creative daughters who, over time, have managed to soften my warrior shell enough to help me understand the importance of sharing this information with others.

Special thanks to Daniel and Shelby Nestor whose guidance and suggestions have helped shape this manuscript for you the reader.

……..and to my inner circle, who will remain anonymous, you have been a driving force from the beginning.

CONTENTS

Introduction		15
Chapter 1	Front Sight Focus	17
Chapter 2	Quitting is Not an Option	27
Chapter 3	Crawl, Walk, Run	37
Chapter 4	Perfect Practice Makes Perfect	45
Chapter 5	Slow is Smooth, Smooth is Fast	51
Chapter 6	Plan your Dive, Dive your Plan	59
Chapter 7	Shoot, Move, and Communicate	71
Chapter 8	Always Have a Swim Buddy	77
Chapter 9	One Force, One Fight	81
Chapter 10	Lead In All Situations	87
Conclusion		93

Warning:

If you are picking up this book thinking you are going to learn SEAL tactics or training regimens, you will be sorely disappointed. This book is about ten phrases that make up a template for achieving mission success. I am going to introduce you to how a Navy SEAL looks at a situation or an overall objective, breaks it down, develops a plan and then sets it in motion. I will show you how thinking like an elite warrior will help you anticipate the next move en route to achieving whatever you are passionate about in life.

    I do have to warn you. This book is not for the faint of heart. Portions of it may even sting a bit to prove a point, but don't worry, the real world stereotypes you every day of the week through advertising and marketing schemes and that wonderful thing they call the internet. Get over it. If you have thin skin thicken it up or walk away. Then again, a good dose of reality may be the wake-up call you need. With that said, you can rest assured the information in this book does not come from a mean place. More than anything I want you to succeed. Achieving your dreams and accomplishing the unimaginable are my goals for you, but I'm not going to sugarcoat the process. This process is not rocket science, but to achieve anything in life you have to start moving forward, and this begins with developing a plan and executing it.

    If you are content where you are, then stay content, but don't read this book; it's just going to aggravate you. This book is about moving forward. If you want something better for yourself and your family, then continue to read. You may even read things in this book you already know. That's okay. I hope the things I write about are more confirmation than revelation.

This book is for those who have been looking for that jump start to help propel them forward in their quest to achieve greatness. This starts by establishing a Front Sight Focus (FSF) mindset.

The information in the following chapters may require you to step outside of your comfort zone. Nothing in life is free. If you want to truly succeed you will have to take a chance at some point. I'm not talking about selling the farm and moving to New Zealand. I'm talking about developing a plan and executing it.

This book is not about ironing out the details for your specific goal; it's about developing a process that puts you on a path towards success. The things you are going to read about in this book aren't based off of five-year case studies, stacks of statistical data, or long drawn-out study groups. This process is about how I have seen and experienced it over the course of 20+ years as a US Navy SEAL.

It doesn't matter what you want to achieve, what you want to be, or where you want to go. This system, this way of thinking, makes it possible if you execute it. With an FSF mindset, you can achieve anything.

The reality is this - if it walks like a duck, and talks like a duck, have a plan in case it is a chicken. I have achieved success because I keep moving forward. I have had my share of setbacks on the battlefield and in life, and have gone to my contingency plan on more than one occasion, but I'm still alive and I am still moving forward and so can you.

I want you to start moving towards your passion in life. I'm rooting for you to achieve your goal. I absolutely want every person to attain whatever he or she wants out of life. I don't care who you are or where you live. I don't care what you look like or how you feel. It doesn't matter. Anyone who puts their mind to something, develops a plausible plan, sets short-term goals and moves forward will achieve.

Introduction

In the Spring of 2015, I was having a conversation with one of my Special Forces buddies. He told me a story about using one of the phrases that he always hears me say, "That's Day One, Week One." He said one of the guys in the office gave him a confused look when he used it. What he had actually said, without that person realizing it, was "You should know better."

As we laughed about the story, the conversation led to how cryptic some military language can be to those who have had limited exposure to that way of life. By the end of the conversation, we came to the conclusion that many of the phrases we use as elite warriors are actually tied to achieving overall mission success, not just in training and combat, but also in everyday life.

Over the next few days, I thought about that conversation a lot. After a short period of time I started writing down some of the phrases that I normally use when I talk to my buddies. I noticed that some started as early as Basic Underwater Demolition/SEAL training (BUD/S). After finishing the list, ten of them really stood out as a template for ensuring mission success, and that is how *Front Sight Focus*, the book, came about.

Before we get started it wouldn't be fair if I didn't fully explain the origins of "Day One, Week One."

When you are learning anything for the first time, the things you learn in the beginning establish the foundation and principles that you build upon as you continue to advance. You could be studying language at your local university, taking an HVAC class at a trade school, or learning to play an instrument. The things you learn on *Day One of the First Week* are the things that are going to stay with you. Think about language training for a second. In the beginning you learn things like numbers, the alphabet and greetings. If you have studied a language you probably still remember normal greetings such as "Good morning," "Good evening," "Good night" and "Thank you." As you advance you transition to more complex vocabulary, verb conjugation and sentence structure. It doesn't matter if it was a long period of time or if it was last month we remember these things because they are the building blocks to which everything is attached.

In the Navy SEAL world *"Day One, Week One,"* things are a little more critical than knowing numbers and colors. "Always have your weapon pointed in a safe direction," "Clear and safe your weapon before leaving the firing line" or "Keep your finger off the trigger until you are ready to shoot" are not only safety issues, they are things we learn on *"Day One, Week One."*

CHAPTER 1

Front Sight Focus

"You miss 100% of the shots you don't take." -*Wayne Gretzky*

"Instead of focusing on how much you can accomplish, focus on how much you can absolutely love what you're doing." -*Leo Babauta*

"*Front Sight Focus*!" In the Spring of 1995, this was the phrase that was embedded in my brain during the marksmanship phase of SEAL Advanced Operator Training. At that time, I was a young frogman preparing for my first deployment as a US Navy SEAL. I don't know how many times I heard that phrase during those two weeks of training, but I do know the Chief Warrant Officer must have yelled it more than 100 times. He did this for a specific reason. He wanted to implant in our subconscious the most important point of performance in marksmanship, *Front Sight Focus*. *Front Sight Focus* ensures that you hit your target with accuracy every time. There are other points of performance that can assist with accuracy, but if the enemy is bearing down upon you and things aren't going exactly as planned, this is the one on which you want to rely.

In defensive shooting, the most important component of sight use on a weapon is to focus the eyes on the front sight. The front sight is the main focal point because, if all else fails, the front sight is pointing to where the gun is aiming.

One of the basic skills of defensive shooting is learning how to focus on your front sight to engage a target. When you shoot paper targets, you have all of the time in the world to adjust your front and rear sights to ensure accuracy. In defensive shooting, there is no time for perfect sight alignment. When you are defensive shooting, you are relying on the balance between sight alignment and point shooting. *Point shooting is shooting at a target without the use of your gun's sights.* This is achieved by focusing on your target, or in the real world, the threat in front of you. As you are focusing on the target, you are simultaneously raising your weapon. Then, as you raise your weapon to engage the target, you transition your focus from the target to the front sight of your weapon. When the front sight comes into focus, you are simultaneously taking the slack out of your trigger and firing the weapon as soon as the front sight meets the target and is crystal clear. When you make the transition from the target to the front sight, the target and rear sight will blur. Using this technique properly ensures success in hitting the target center mass.

Building upon this technique, a person can use this concept to develop a *Front Sight Focus* (FSF) mindset. Developing an FSF mindset requires an "All In" mentality. Not only believing you can achieve success, but visualizing it as well.

To develop an FSF mindset, you have to make the initial decision to move forward. It may be deciding to make the transition from an amateur to professional status. It may be taking the next step in accomplishing a goal such as a degree or certification in your occupation or field. It may be stepping up your game for the next promotion, such as dressing more professionally or arriving for work 10 minutes early instead of racing the clock to get to your workstation before anyone notices. Whatever the reason, this is where you have to decide to what level you are planning to move. You need to really sit back and think about this and then decide your plan of action. You need to allow this to really soak into your brain housing group. This decision is where your mission-planning process for life begins. Is this a short-term mission or a long-term mission? Are you all in at all costs? It's all about what you want to achieve.

Everything is achievable, but there are a number of variables that play into the planning process of your mission. Depending on the scope of your objective, those variables could range from time and financial, to skill level, your current location or a bit more of well-placed effort. "Easy day." All you have to do is identify the obstacles and develop a plan to establish a solution. Too easy!

What's your first response when you read those two words, "too easy"? "Easier said than done?" Am I close? I didn't use those two words to demean you, your mission, or your overall objective. I used those two words for only one reason, to show you how I think about everything. I look at everything as achievable, a "You can't hurt me" mindset. I will figure out a solution, "too easy." From this point on you have to think like this, talk like this, plan like this, expect like this……everything like this. *Did you know that there are companies starting projects even though the overall objective isn't even physically possible at the present time?*

Their thinking (mindset) is if we keep moving forward towards the goal we will figure it out. We will make the discovery that makes it possible.

If you are ready to move forward, you need to identify your starting point. This is done by completing a realistic assessment of your skills/experience. You have to be honest. It's you we're talking about, not somebody else. It's you! You are making an assessment of yourself. You are the only one who knows you. You know your realistic capabilities. This is not about what someone else stated or implied that you were or were not capable of accomplishing. You can accomplish anything you set your mind to, but at this point in the game, it's about identifying and establishing what you possess to start the mission planning process. What skills/equipment do you bring to the table? I will even go you one better. I don't care what your capability is. This can be overcome, but you have to be honest. This information is for realistically setting your starting point, which in turn helps develop your timeline and identify your short term goals. Keep in mind the majority of goals in life aren't achieved overnight.

What I am about to explain in the following chapters is how this mission planning process works within the SEAL Teams and how you can adapt these same principles in real life. In the SEAL Teams, our mission planning process doesn't start after you graduate BUD/S training. The process starts on "Day One, Week One" of BUD/S training. Graduation is one of many short term goals en route to becoming an elite warrior. The day you identify your overall mission objective (what it is you want to accomplish) and take that first step forward is your *"Day One, Week One."*

It always amazes me how accepting people are when it comes to believing only specially gifted people achieve unimaginable success. I call this life propaganda. It's the same negative mentality of underachievement that is passed down from father to son and mother to daughter, generation after generation. It's exactly that - "crap." You have the capability to achieve anything you want to. I mean that, but it does involve doing a current, honest, and realistic capabilities assessment.

Self-assessment is one of the hardest things in the world. It's not actually hard, it's just very uncomfortable. Deep down we already know our own shortcomings. We hate being confronted about them for that very reason. Most people don't ask the hard questions because that's what makes it real. It brings our shortcomings to the surface. Nobody wants to hear that kind of truth coming out of their own mouth. Take a look in the mirror. Get buck naked and look in the mirror. The question is simple. "Do I like what I see?" If you don't like what you see, then chances are you are not doing what you like to do. When you are doing what you truly love to do, your fitness and health will change. Everything about you will change. The way you talk and dress will change, and you will start exuding confidence. When you are moving forward toward something you are truly passionate about, you are going to like who you are and it is going to be very apparent to those around you.

Your situation in life (status, education, skill level, financial) will directly affect the rate at which you achieve these short-term goals. If your goal is to be a teacher, you cannot expect to achieve that until you achieve obtaining a Bachelor's degree from an accredited university. The achievement of the short-term goals will then affect your longer term objective, such as becoming a high school principal, for example. When you decide to pursue what makes you happy, time really doesn't matter. When you make that decision to move forward in the process of achieving something you are passionate about, it's no longer work. It's going to be what drives you to succeed.

If you are serious about pursing life with an FSF mindset, you may have to get rid of a few things. If you possess any of the following items, it is not possible to continue until you rid yourself of them:

Item #1 Excuse Machine – There is no room for excuses in the FSF world. None. This doesn't mean mistakes won't be made along the way. Successful people make mistakes all the time, but they take responsibility, figure out a solution and keep moving forward.

Item #2 Blame Thrower – Most people who own an excuse machine eventually acquire a blame thrower in an attempt to shift the focus to someone else. This is standard operating procedure when excuses no longer work.

Item #3 Fault Finder Virus – This is a sickness that infects the whole mind. It corrupts and skews a person's view of everyone and everything. It normally starts with an inability to ever admit being wrong….about anything, and rapidly spreads from there. It is easy to spot a person infected with this virus because of the negativity that oozes from everything they say or produce. Once consumed with this virus, a person finds an overwhelming desire to undermine everyone and find fault in everything. *If you ever come in contact with a person that is infected with this virus, walk away. Do not engage in conversation with these types of people. People infected with this virus need to seek professional help.*

There is not enough time in life to engage with people who are filled with excuses, like to find fault, or blame others for their inadequacies.

In order to move forward, you have to cut away all the outside collateral waste. You need to stop worrying about the other guy or that other woman. You need to concentrate on your own short-term goals that are going to move you forward toward your overall objective. That includes stop hating people who have more than you. It doesn't matter whether they deserve their position or the things they possess. You have to stop dwelling on what everybody else is doing or acquiring. Most of the time you don't know their history or how they acquired X. Put that energy into yourself and your movement forward to achieving your short-term goals en route to your terminal objective.

Never compare yourself with a celebrity. Do not allow yourself to get caught up in how they live or what they drive. Half of that stuff is scripted, staged or made to look a certain way. Stop trying to keep up with the Joneses. That's a complete waste of energy. That energy could be used in a positive way. You should use these things as motivation and inspiration, not jealousy. Time is so precious. Use these things to your benefit. It is not up to you whether somebody deserves something. Hate and jealousy will get you nowhere. Jealousy is like a major malfunction on your main parachute. You have to cut it away and go to your reserve because it is not going to get any better. It is counter-productive and one of the main ingredients needed to create self-sabotage. Jealousy is a non-starter.

Now back to where I stated, "Everything is achievable." I believe that statement with all my heart, but that "everything" portion of the statement needs to be taken with a good helping of reality and a scoop of common sense. A good example of this is professional sports. You could tell me your objective is to be a lineman for the Dallas Cowboys, but you are 53 years old, 5'6" and weigh 157 pounds. I would tell you right from the start your front sight is out of focus. There are things in life that are difficult to overcome and most of those have to do with physical characteristics such as age (time), size (height or weight), speed etc. If physical characteristics are limiting or hindering your overall goal, step back, take a deep breath and re-adjust your sight picture. In the professional sports realm such as the National Football League (NFL), becoming the oldest and smallest starting lineman for the Dallas Cowboys may be a stretch too far, but being a part of the Dallas Cowboys organization is not. Working in the front office, or even coaching the linemen in some capacity is definitely possible and has a better chance at longevity than being an actual starting lineman. In the same scenario, if age isn't a limiting factor, then there are at least two positions on the Dallas Cowboys football team that have nothing to do with age, size, or speed if you can punt or kick the football farther than anyone else.

Sometimes you just need to scan the operational area and engage the targets (potential goals) one at a time until you find the one that is going to be your obsession, the one that drives you day in and day out. If becoming a professional athlete is your objective and longevity is the goal, then professional golf may be an option. The point is this. Shoot for the stars. You can aim as high as you want, at anything you want and adjust your fire from there. Sometimes you will need to walk the target in until you hit the black, the bullseye, and figure out what you are truly passionate about, your overall objective. Some things in life look glamorous from the outside, but the blood, sweat and tears involved to reach that objective may not be for you. The identification process and the journey are different for everyone.

Age may control certain parameters or choices, but in some circumstances it may enhance your ability to easily identify your overall objective. Age or life experience may even propel you to achieve your goals at a faster rate. A younger person may have more opportunities ahead of him or her, but the learning process may take longer due to the "life" things he/she still needs to acquire en route to achieving the overall goal. In most circumstances, experience requires time.

It's not about trying to achieve the biggest thing possible. It's about identifying what you are passionate about, what drives you, what makes you love to get out of bed every morning. If you focus your front sight on achieving that, greatness is right around the corner.

CHAPTER 2

Quitting is not an option.

"If you think you can or you think you can't, you are right."-Henry Ford

"Winners never quit and quitters never win" –Vince Lombardi

In the SEAL teams, if you are about to be critiqued on an event that you have just finished, and it didn't go as well as it should have, you might hear your debrief prefaced with the phrase "Take off your thin skin." This is team guy code for put down your egos, listen up and let this information soak into that stubborn muscle between your ears. "Now take off your thin skin and listen up!"

I originally had this phrase earmarked in my outline for one of the chapters at the end of the book. The plan was to tie it all together with a "Quitting is not an option" speech. Sort of like a go forth and do good things, don't ever look back, keep moving forward mantra, but the more I thought about it, I decided it has to be in the beginning.

You have to believe that any type of negativity is a non-starter in all mission plans designed to achieve any type of success in life. An "I Can't" mindset leads to a "We Can't" mindset. You don't have time for it. You don't need it. You don't want it and you have to tell yourself you will not tolerate it. If you are one of those people who brings that weak-minded mentality to the game, just put this book down and go home. Thanks, but no thanks for coming. I don't have time for quitter anything. Negativity, whining and excuses go nowhere. If it isn't working out like you planned it, tough. Finish the evolution, event or project first. Then re-assess the situation, come up with a solution, and re-engage the target.

It may not always be a sunny day. There are going to be times where you will get frustrated, but never say "I can't get this." This is about changing your mindset and your language. Use phrases like "I'm having difficulty with this" or "This is challenging." The difference is the word "can't." "Can't" is the key ingredient in a defeatist mindset. Changing the language to "I'm having difficulty" allows you to vent and express frustration, but also establishes the "I'm going to get this, I will continue moving forward, I am going to win" way of thinking. It's as simple as that.

When it comes to quitting, don't do it! Don't entertain thoughts about it. Don't make it one of the options when the going gets tough. I moved this phrase to the beginning of the book because that is where it is introduced in BUD/S training. Right from day one, you are informed of your options to take when the going gets tough. You can "Quit" or you can "Suck it up" and move forward. If you choose to quit, the instructions are quite simple. March over to the bell that hangs on the pole outside of the first phase office. Stand next to the bell on top of the two frog feet that are painted on the ground. Take off your phase helmet and hold it in your left hand as you ring the bell for EVERYONE at BUD/S to hear. State your rank and name and say "I Quit." As soon as you do this, someone will assist you with filling out your Drop On Request

(DOR) paperwork and you can be on your way to fulfilling your U.S. Navy dreams somewhere else.

This process may sound a little harsh, but the process for earning the right to wear the Navy SEAL Trident on your chest is not part of the kinder, gentler, everyone gets a trophy process that society is busy trying to assimilate you into. It is a warrior process that cannot be stomached by most. The fact of the matter is, being a Navy SEAL is not for everyone, but going through life with an "I can achieve anything I set my mind to" attitude is for everyone if you want it badly enough.

A lot of people think Navy SEAL candidates quit BUD/S training because of the physical demands and requirements. That couldn't be any further from the truth. The reality is US Navy SEAL training is more mental than it is physical. Navy SEAL training gets a lot of exposure in the media through news outlets, documentaries, books, movies and the plethora of information that is floating around on the internet. Most of this hype plays to the physical demands of the training such as the extremely long swims and runs, Hell Week (extended sleep deprivation), Drown Proofing, Log PT, Night dives; the list goes on and on.

Having had the full benefit of this training, I can tell you it is a schedule of physical activity that starts from the time you get out of bed to the time you get off that night or the next day. At times it can get overwhelming……. if you let it. I know sometimes that can be easier said than done, but you have to look at feats of this magnitude through the Front Sight Focus mindset filter. Nothing is impossible. *A year and a half prior to going to BUD/S training I couldn't swim the full length of an Olympic size pool without stopping to catch my breath.* Through continuous work and a can do attitude, I was able to slowly increase my speed and endurance in the water.

You can do anything you set your mind to. When you go to do something that others have already achieved, quitting should never enter your mind. If you aren't the first person to ever be in that situation, attempt that feat, go through that training, and succeed, then quitting is not an option. If it has already been done, achieved, or completed, it's possible. Running a four minute mile was, at one time, considered "impossible" until Roger Bannister first accomplished it in 1954. Now, if you don't run a four minute mile, you are laughed off the track!

One of the things my classmates and I would tell each other during BUD/S training was "Let's just get to chow." Get to the next meal. We knew they had to feed us. During regular training, it was mandatory that they feed us three times a day. During Hell Week, we ate four times a day. We knew that every six hours during Hell Week, there would be a mandatory chow time. During this time, breakfast, lunch, dinner and mid-rats (Midnight Rations) was our reset. We concentrated on making it to chow, making it to our reset point. We knew once we got to chow, we had an allotted amount of time to rest, get hydrated, get refueled, and most of all, get motivated. This type of motivation was a little different. It was more along the lines of camaraderie. It was laughing about the struggles and shortfalls that we encountered during the last evolution. This time motivated us, built confidence and helped establish that "Been there, done that," "You can't hurt us" attitude.

Regular life is no different from the Hell Week scenario. You have to figure out what your everyday reset is and concentrate on reaching it. It may be your meal times. It may be the time between when you drop the kids off for school, soccer or dance and pick them up. Whatever it is, it has to be a regular occurring event to help establish that drive to get there. Some people don't need a daily, or even a weekly reset point to catch their breath or get caught up, but if you do find yourself in a high stress situation, don't let it get out of hand. Break it down early. Identify reset points and short-term goals, then concentrate on reaching those milestones.

To build unity and motivate each other during painful times at BUD/S we would sing. Our song of choice happened to be Jimmy Buffet's "Margaritaville." As soon as we locked arms as a class during surf torture evolutions, or were all lined up in the push-up position during a hammer session, you were guaranteed to hear "Margaritaville" over and over. Each time, it would get louder and louder as we motivated each other. This helped take our focus off of the horrible situation we were enduring at that time. When the going gets tough, find your "Margaritaville." It may not be a song. It may be something you think about or visualize.

At no point in life is quitting an option. I remember when I was a young boy my dad told me I would play at least one year of little league baseball and one year of football. No questions asked. I didn't have a choice. He did say though, that after one year, if I didn't like it, I didn't ever have to play again. So, throughout the season in both sports, I never even thought about quitting, no matter how bad or tough it got. I knew quitting was not an option right from the start, so I just concentrated on playing the game. At the end of the season, the sense of accomplishment built my confidence to take on larger adventures. You have to establish right from the start that quitting is not an option. Finish the evolution first, then decide whether you are going to continue along this path. Some things in life are not for everyone, and sometimes you don't know this until you try it, but don't quit halfway through because the going gets tough.

Don't give yourself, your kids or your family the option to quit. This life phrase needs to be non-negotiable. Quitting is a virus. Do not let it enter your mind. If you quit once, it becomes easier and easier to quit each time and it can become contagious and go viral. Quitting can spread throughout your family and into your work place. Once you are infected with quitting, it becomes harder and harder each time to prevent that "quitter mindset" from taking control when the going gets tough. If you don't get it under control, it leads to the worst kind of mindset, a "negative defeatist mindset." A negative mindset expects failure. It searches for it at every turn.

When you break it down into its simplest form, most people don't want to quit. The majority of people don't go out looking to quit. When attempting anything in life, you have to keep moving forward if you are going to achieve any type of goal, no matter the size. Perseverance is a life tool a lot of people are missing in their personal tool kit. I think the word itself sometimes scares people. All you have to do is remind yourself to keep moving forward. Succeeding is all about eliminating that "give up" mindset. The reality is most of the time people quit or give up because they take on too much too quickly.

I can't emphasize this enough. Most of the time, taking on too much too quickly is a recipe for disaster. Before all of you naysayers gang up on me, hear me out. My too much too quickly is different from your too much too quickly and his/her too much too quickly and so on and so forth. Everyone has his or her own limits. Know your capabilities, know where you are in the process and know your experience level. You need to be honest with yourself in everything you do. This is not a scare tactic; it is a warning order for what is eventually going to happen.

When you start achieving short-term goals, your confidence will increase. The way you talk will become more polished and your body language will exude it. This is where a positive mindset, a keep moving forward attitude, takes you. Your mind will start believing anything is possible and your subconscious will just naturally fall in line with the mind and start working on a solution to everything you do. This is the avalanche effect of success that you should always be striving to achieve. Now, if you immediately shift to an "all in" mentality right from the word go, success can come quickly. So I'm here to give you this advice. If it does happen quickly, always remember "Slow is smooth and smooth is fast." *(I'll explain this phrase in chapter five)* Knock down those short-term targets one at a time. Transitioning too rapidly between targets or "taking on too much too quickly" can cause extra bullets to be discharged at a target/goal when it could have been knocked down just as quickly and easily by just slowing down and executing the basic fundamentals/principles that have gotten you to this point.

Think about all of the people you know who have made New Year's resolutions to lose weight or get into shape. I don't know the exact statistics, but I bet gym memberships increase every January. Then, as the going gets tough, most of the new members fade away into the shadows. When you break it down, I'll bet the majority of people who give up right away try to take on "Too much too quickly." It is very difficult to go from your current lifestyle or current eating habits with minimal exercise to an all-out workout program overnight. It's not healthy for your mind or your body. The psychological effect alone will have you cancelling that gym membership in no time. You have to have a plan that allows the whole you, your mind and body, to make the transition. *Cold turkey really only works when the fear of something is involved.* In the "getting into shape" scenario, controlling your eating habits first is a better place to start. I would suggest reading *It Starts with Food* by Dallas & Melissa Hartwig to educate yourself about how the different types of food affect your body and move on from there. Getting into shape is a lifestyle change that consists of much more than joining a gym and starting a workout routine. You need to have a plan for making changes and transitioning into the new lifestyle. You need to be able to successfully manage areas such as work, family, rest, food intake, and workout if you want to achieve long term results.

Don't become fixated on the overall objective and the amount of time it may take to achieve it. Setting short-term goals, which may sometimes be baby steps, is the only way to move forward.

You need to knock down the targets in front of you. Focus on the short-term goals. This may just be making it from one evolution to the next, getting to lunch or making it through the day. If you are moving forward, you are on your way. Just keep moving towards the next target. This in itself, tells your brain and the subconscious you are serious about the plan. You are serious about your future. You don't have to make your first million off of your first idea or product that rolls off of the assembly line. Sometimes it is a journey to the terminal objective. There is nothing wrong with that as long as you keep moving forward and don't let it overwhelm you.

CHAPTER 3

Crawl, Walk, Run

"It's not the will to win that matters- everyone has that. It's the will to prepare to win that matters." –Paul "Bear" Bryant

This phrase is a guide to navigate you and/or your team as you develop a solid training plan for achieving your objective. When you are looking to master a technique, a method, or a skill, it is imperative that you use the *Crawl, Walk, Run* process. Using this approach to design a training plan ensures you are building an efficient system tailored to you so that you can maximize your time and progress. *Crawl, Walk, Run* isn't a catch phrase or the latest and greatest training trend. It is the pipeline to ensure greatness is accomplished at every level along your route to achieving your overall goal. There is one caveat. When it is executed correctly, it involves more time than most people want to endure. If executed correctly, it ensures that greatness can be achieved. Remember nobody rides for free.

When developing a fundamentally sound *Crawl, Walk, Run* training plan you need to ensure you use the "Keep It Simple SEAL" (KISS) principle. Don't get caught up in the word "simple." Some of the plans needed to master certain critical skills are going to be anything but simple. The emphasis of this principle is actually implied. What it is really saying is use "common sense" when you develop a plan. The KISS acronym was designed to remind you to not make it so complicated that you invite failure, but sound enough to achieve the desired result. When designing your plan, you need to know your actual realistic capabilities and if you are working as a team, this also includes those of your teammates. The main goal should always be to prevent "Murphy" and his family from having any reason to show up to your situation or event. Remember *Murphy's Law: "Anything that can go wrong will go wrong."* This is accomplished by executing good sound principles and keeping the "Good Idea Fairy" out of your planning process.

Crawl, Walk, Run is designed to control the speed at which you are learning to ensure that you master that technique or skill level before moving to the next step. At each step of the process, there are specific requirements that have to be met. Depending on what type of training it is, the introduction of stress at various intervals is paramount to evaluate true learning. Testing is added when certain learning sets require specific scores, times, or criteria to be met in order to move on to the next step. When certifications or qualifications are part of the overall objective, testing helps identify any deficiencies, flaws or areas that need remediation. Crawl, Walk, Run is a system that is used to ensure that the outcome is true learning and that muscle memory is being developed at every level along the route to meeting your objective.

It doesn't matter whether you are studying to be a doctor, learning to swing a golf club or training to be a US Navy SEAL. This process is designed to cover all the bases so that the things you learn and the techniques you develop are embedded in that brain housing group that sits atop your shoulders. This process ensures the techniques and fundamentals are woven into your muscle memory so they become second nature.

The primary responsibility of the *Crawl, Walk, Run* process is to maximize learning and development through the use of best practices. To be successful and maximize your time and progress you need to learn from the best and this needs to be done right from the start.

A concept I use when developing any type of plan is the Educate, Train, Anticipate (ETA) Concept. To begin the development of a sound training plan, you need to *research* and *educate* yourself on the best practices available. This is done by identifying the best learning institutions, talking to the experts, attending seminars, reading books, watching videos and calculating the costs involved. This requires you to get smart on what it is that you need to learn or develop in order to accomplish your goal. After you have exhausted all avenues in the *Research* and *Education* phase, it's time to take that information and put it to work. Meet with the teaching professionals in the field you are planning to join, enroll or schedule your first class or lesson and establish your *"Day One, Week One."* I strongly suggest going to the best institutions to acquire that diploma or certification if it's knowledge-based. If it's a technique or method- type skill "TAKE LESSONS FROM A PROFESSIONAL!" Learn what "right" looks and feels like before you venture out on your own to practice. Perfect practice is only achieved by knowing how to execute "right" the most efficient and effective way. Build that muscle memory by executing perfect practice over and over again.

Let me give you a little taste of what is involved with the typical "Crawl" phase of this process when starting weapons training in the SEAL teams.

During the third phase of BUD/S, the students begin small arms weapons training at the "Island" *(San Clemente Island)*. The initial Crawl phase starts with an "Education" of each of the small arms used in the SEAL teams. In a classroom environment, the students are taught the specifications of each weapon and the various types of ammunition they use, how they function and how to clean, maintain, and repair each of them. They are taught firearm safety, how to load, unload and make each of the various weapons safe. The students are tested on each of the weapons for knowledge, along with a practical test, requiring them to break down each of the weapons and reassemble them for time, and then tested again with a blindfold.

At the "Island," the BUD/S students spend a large part of their time on the weapons range, working on the basic fundamentals of marksmanship, developing good shooting mechanics, sight alignment and accuracy. This is done to ensure each student is at least a Sharpshooter, with the majority of the students graduating as Expert marksmen. As the students progress through the phase, they learn to shoot and move in pairs and then move on to engaging targets as a squad. Small unit tactics are then introduced using various types of terrain to work on patrolling, communications and fields of fire. At every point throughout the *Crawl* phase of weapons training, the training builds upon itself from evolution to evolution. Rehearsals and walkthroughs are always conducted and debriefs are used to fine-tune shortfalls from each training session. But here is the key to why this *Crawl* training plan is efficient and effective; the instructors are the best of the best, and were taught and trained by the best shooters in the world, using the best practices that have been proven to work on the battlefield. Who are you going to learn from to operate on "your" battlefield?

The *Crawl* phase of weapons training doesn't end with graduating BUD/S. It continues on into SEAL Qualification Training (SQT) where students are introduced, educated, trained and tested on rocket launchers and foreign weapons. During SQT, the pace picks up and students start to *Walk* as stress and time are added to the equation, the weight of the shooters equipment increases, and night vision equipment is used more frequently.

In the SEAL Teams, this *Crawl* process is used for every type of training that is conducted for the first time. The purpose of the *Crawl* phase is to produce those core building blocks needed to build the foundation that is being engineered throughout this process.

Once you have educated and trained yourself to *Walk*, it is time to start picking up the pace. This is where you take that certificate or diploma and go get some experience. If your overall objective is to become a professional in the medical field, practicing law or working in the financial markets, then your "On the Job Training" (OJT) is probably going to consist of interning at an established institution, firm or company. If it is a technique, method or specific trade-type skill, this is where stress and testing are added to dial in times, scores or specific selection criteria that is needed to achieve the required results (short-term goals) en route to your overall objective.

In the SEAL Teams, the operators are off to a fast walk upon graduation of SQT, where they are pinned with a Navy SEAL Trident as they head off to their respective Teams. At this point, the SEAL operators transition from a fast *Walk* to a jog as they join their Team. On the Team, the operators are assigned specific jobs or positions within the Task Unit or Platoon and the advanced training continues. The SEAL operators begin to *Run* as they enter unit level training and learn the Team's Standard Operating Procedures (SOPs). As the operators gain experience and continue to excel, the Team sends them to specialized schools to acquire specific qualifications that will continue to elevate their skill level.

True running only occurs with experience. Experience is the only thing that prepares you for the unknown, the stressful situations, contingencies, emergencies, disasters etc. You can practice till you are blue in the face, but "real world" on-the-job experience is the true test, that final exam, that validates you as a professional, as the go-to expert in your field or position.

You see this all the time in sports. You can have extremely talented athletes melt down or fall apart against lesser opponents, when faced with an unknown or stressful situation that is being experienced for the first time. The chance of implosion, from the lack of experience, goes up exponentially when you add a timeline and multiple player responsibilities. There is some truth to the quote "You are only as fast as the slowest member on the team." Keep in mind, the word "slowest" refers both to the physical and mental aspects of those team members. When success is dependent upon the entire team, experience needs to spread across the whole spectrum of its advertised capabilities. This is why there are conditions and standards required for certain jobs that rely heavily on experience to ensure mission success.

Once you have some experience under your belt you can start to use this to your advantage when situations or opportunities present themselves. The more experience you have the easier it is to "anticipate" the next move. The *Anticipate* portion of the ETA Concept is the advanced phase of the *Run* process. Once you have mastered the ability to "anticipate" the next move, you can begin to side step life's potential hazards, predict opportunities, game plan effectively and win! In this mindset, game-winning translates into the accomplishment of your short-term goals en route to the achievement of your overall objective.

Initially your inspiration, passion or interest in something may have been generated by watching or reading about someone else's feats or accomplishments. As you move from short-term goal to short-term goal while graduating through the *Crawl, Walk, Run* process, your adventures, experiences, and achievements take over as your inspiration to continue the quest. You and your accomplishments become your own motivation. Changes in your body, if health and fitness is the overall goal, your abilities and production, if the overall goal is skill-based, knowledge, test results and ranking if it is education-based, and so forth. There is nothing better in this game than the transition of your initial inspiration to becoming your own inspiration as a direct result of your own achievements. This transition will boost your motivation and confidence by leaps and bounds as you continue to gain momentum in your pursuit of mission completion.

If you have identified what you are passionate about and have taken on an FSF mindset, start moving towards those 10,000 hours that Malcom Gladwell talks about in his book *Outliers,* the ones that lead to greatness.

Always remember what the difference is between an Amateur and a Professional. An Amateur practices until he gets it right; a Professional practices until he never gets it wrong.

CHAPTER 4

Perfect Practice Makes Perfect.

"He who loves practice without theory is like the sailor who boards ship without a rudder and compass and never knows where he may cast." -*Leonardo da Vinci*

Now that you have identified what you want, done an honest assessment and understand the significance of the *Crawl, Walk, Run* process, we need to integrate the "Perfect practice makes perfect" principle.

Nobody is perfect. I beg to differ. Perfect is how you define it. In certain situations, perfect may only be for a split second. It could be making a game-winning shot or taking that once-in-a-lifetime photograph. Perfect doesn't have to be forever and it definitely is not going to last your entire life, but it doesn't have to. The kind of perfect that I am talking about is perfect for a specific reason, situation or event.

Let's take the sniper situation for example. The sniper practices his perfect mechanics for shooting over and over and over. He practices reading wind, breathing, trigger squeeze, positioning, concealment and shooting stationary and moving targets at known and unknown distances. The sniper does this so that when the day comes, he is perfect for the reason, situation or event. *"What is perfect in the sniper realm?"* In shooting competitions, it is very apparent what is considered perfect; Bullets in the black, the bullseye or a certain score. In the combat environment, perfect is the successful protection of an individual, an assault team, the patrol, the convoy or the force. The key word is "successful."

    The same is true back in the real world. The same rules apply. A true professional basketball player puts hours and hours into perfecting his/her game. He or she shoots from all different locations on the basketball court, using the best shooting mechanics to fine tune accuracy. Shots are practiced in the rhythm of taking a pass from a teammate, shooting off of the dribble and shooting from all different types of defensive coverage. The player practices while being guarded man-to-man, zone or while being double-teamed. A player repeats these shooting drills so that he or she can have the best possible chance for success on the court when the game is on the line for their team. Just like in the sniper business, a basketball player's perfection can be scored or evaluated by how many shots go in the basket during a shooting competition. In reality, the player's true perfection is only measured by whether the team was successful in winning the game or whether he or she made that game-winning shot. If a player makes every one of his or her shots during the game, but the team loses, there is nothing perfect about the player, the team or the organization.

    Don't take the phrase at its literal meaning. It is worth far more. This phrase is a way of thinking. It's a mindset. It's an objective that gives you the desire to achieve success... unimaginable success, but there is a caveat. It's not the fact that you have the desire to practice or that you are obsessed with practicing until you get it right. It has to be more than that. To be truly successful, it has to be perfect practice.

Perfect practice is practicing using the best techniques, mechanics and trainers. Once you have integrated the best practices, you can then couple it with the element of desire, that true passion to be the best. The well-known phrase that I mentioned at the end of chapter three says it best, "Amateurs practice until they get it right. Professionals practice until they never get it wrong." This can only be achieved if you are executing perfect practice. Some things you can practice your entire life, but if it is not the best, most efficient way of practicing, it is just going to bring you short of true success.

There may even be variables involved. Most of those variables have to do with the wonderful world of physics. The reality is the sniper is only capable of shooting as well as the gun will allow. This is why elite warriors have all types of weapons for specific situations. Snipers predominately use longer-range weapons than assaulters, who are normally in tighter, more confined spaces. Assaulters normally don't need to shoot extreme distances because the enemy is right in front of them. So before you start hunting down the best practices or weapons, identify what type of "shooting" is going to be required to achieve your assigned tasks.

Golf is a good example in the real world. Each club in the bag requires a slightly different technique to be used for certain situations during a golf game. Drivers are used for driving the ball longer distances. Regular irons are used for varying distances on the fairway. Wedges and higher-numbered irons are used for the short game in and around the greens. The putter is used for putting the ball on the greens. There are trainers who specialize in one or more of these situations that are encountered while playing golf. The same thing works for shooting. You have expert pistol shooters, experts in rifle shooting and experts that work on basic shooting mechanics.

The key is identifying what portion of your game you are looking to perfect. I use the word "game" as a general term for whatever it is that you are interested in establishing perfection. Sales communication could be your game. If it is, hunt down the best sales books, seminars, tutors etc. Don't just go to one seminar or read one book. Explore and do your own research. The goal is to be the best, so learn from the best.

The SEAL teams don't just go to one shooting professional to learn all of the various mechanics and shooting techniques. We hunt down the best shooters in the world with the most experience. We do this so we can gain the best knowledge to establish the latest and greatest perfect practices. The game is constantly changing and always getting better. Someone out there is always bringing something better to the table.

Take a look at professional sports. There are companies specifically designed to prepare upcoming professional athletes to excel at things like the NFL combine or special drill days that certain athletes get invited to prior to the draft. People are discovering better, faster, more efficient ways of doing things to give them a better competitive edge over the competition. You have to be constantly striving to stay ahead of the game by doing research in the field in which you want to achieve success. The key is never resting on what you have achieved in the past. The day after you have been successful at achieving a short term goal, it instantly becomes the past. You are only as good as your last job, shot or presentation. You always need to keep moving forward.

In chapter one, I talked about *Front Sight Focus* being so important in defensive shooting, but just knowing that information isn't going to get it done. That technique needs to be practiced over and over again to build muscle memory. That muscle memory is not just physical muscle. It has to be implanted in that mental muscle between your ears. In defensive shooting, your technique needs to be second nature, no actual thought needed when you draw that weapon from its resting place. That movement needs to be one natural fluid movement from start to finish, so when perfection is needed for that specific reason, situation or event, your training takes over to ensure mission success. In our world as elite warriors, if we are shooting our pistols, there is a high probability that our primary weapon is down or it is out of ammunition. Either way, the stakes are high and there is no time for mistakes.

To ensure this muscle memory has been embedded, we implement all types of stressful situations into our training. We integrate physical stress such as sprints, push-ups, pulls-ups, dips, all types of physical activity to elevate the heart rate and produce that adrenaline rush. Then the same drills are run for time and accuracy over and over, then again in competition with our peers. The training continues with the implementation of various equipment loadouts, vehicles and element sizes. The training then transitions to night operations, and the process is repeated.

Hunters do the same thing when training to hunt turkey, deer, or big game. They want to put themselves in the best possible position in order to successfully accomplish their goal, which is taking that animal. They have, through trial and error and repeated practice, gained knowledge in order to choose their hunting area, their clothing to compensate for the weather and odor elimination. They have chosen their weapon and ammunition for the anticipated distance, whether there is any brush, and the size and knockdown power needed. They have practiced getting into their area or tree stand in the dark numerous times so that they don't waste valuable time in the morning in reaching the hunting ground. They have spent numerous hours shooting, practicing their breathing, practicing the use of their game calls and getting their mid-day sandwich out of their packs without making noise. They simulate the hours and the specific moment right before they take the shot. They want to be sure they can handle the rush of adrenaline when they see that once in a lifetime trophy 12-point buck walk out of the wood line. They want to ensure that, even with the adrenaline rush and the high heart rate, they can maintain their accuracy.

Believe it or not, most of the time the thing that separates you and that person ahead of you is technique. That person has a better technique, a better system, a more efficient way of doing things because they have more knowledge of that objective that you seek.

CHAPTER 5

Slow is Smooth. Smooth is Fast.

"A particular shot or way of moving the ball can be a player's personal signature, but efficiency of performance is what wins the game for the team." –*Pat Riley*

One of the first things you quickly learn in the SEAL teams is the phrase "Slow is Smooth, Smooth is fast." The first time you hear this, it may not sound like it makes much sense, but what it actually is saying is "Slow down, relax, and let your training work for you." Everything we do in the SEAL teams has a technique, a mechanic or a procedure that has been developed over a period of time and perfected using the best practices available from some of the greatest minds in the game. We train using these techniques in all environments and situations, using all types of equipment, day and night, with and without stress so that when the time comes, your actions will be second nature and purposeful. The time we put into developing these motor skills prevents the brain housing group from going into a spastic, out of control, "running out of time" state of mind when an adrenaline rush is brought on by a stress-induced timeline. When you incorporate the "Slow is smooth, Smooth is fast" mindset, you are literally removing stress from the situation. You are being deliberate with your thinking, your movement and your decision making process. You are making it one fluid movement. This is the key to navigating any type of stressful situation with success.

Now there is a catch to this process, which I alluded to about halfway through the opening paragraph, when I said *"We train using these techniques in all types of environments and situations, using all types of equipment, day and night, with and without stress so that when the time comes your actions will be second nature and purposeful."* This training process plays a major role if success is to be achieved when stress is encountered. You have to be knowledgeable about the subject, situation or event. To overcome stress, you have to adequately prepare yourself for its eventual arrival. You have to properly prepare yourself for it, so when it comes, you can identify it early and take the required steps to successfully navigate through it. That is what allows you to adapt a "Slow is smooth, Smooth is fast" way of operating.

During shooting evolutions "Slow is smooth, Smooth is fast" is often heard as a constant reminder as the various drills and best practices are executed. This is done to remind the operator to be deliberate and purposeful with his movement and decision making process, but also to use the mechanics that he has been taught so that he builds good muscle memory.

In training, missed shots lead to running out of ammunition early, which can easily be fixed by walking off the range and reloading. On the battlefield, during a gunfight, this is not an option. You only have what you are carrying, and when the magazines are empty, you don't get a "Mulligan" or a "Training Timeout." We train accordingly, so that every shot counts, has purpose, no matter whether you are transitioning from your primary weapon to your secondary, moving from one obstacle to another or using night vision goggles in times of limited visibility.

That is why all of our preparation for the battlefield includes stress factors in every facet and at every level of our training pipeline. It doesn't matter if it is something as simple as giving an IV during a down man drill, our training process properly prepares you. In the *Crawl, Walk, Run* learning process, IV training is reintroduced at various intervals throughout the training pipeline at times of high stress, which allows the operator to build confidence and *experience*. This helps foster the development of the "Slow is smooth, Smooth is fast" principle.

Your typical IV training progression may look something like this: The training would start out with educating the operator on best practices of giving an IV to a victim and allowing the operator to practice and master this skill in a sterile classroom environment. From there, stress factors would be introduced throughout training and at every level. Those stress factors could include performing IV training following evolutions that involved raised heart rate, in the back of a vehicle on a bumpy road, at night, or in contact drills during urban combat training. *It is not uncommon to see IV training at the most inopportune times throughout our training. This is to prepare you for the unexpected.*

One of the early evolutions in BUD/S that brings this concept to light is the underwater knot tying test following Hell Week. The actual event is quite simple and the instructions are relatively easy to follow. When the event is coupled with some instructor-induced stress and any ailments you may have brought with you from the previous week's adventure (Hell week) such as micro lungs from congestion or colds, ear infections etc., your ability to hold your breath, clear your ears and conduct the assigned tasks can become challenging.

To pass the test you are required to tie five knots: Square knot, Beckett's Bend, Bowline, Clove Hitch and a Right Angle. These knots are individually tied onto a line of rope called the "Trunk Line" that is stretched across the pool at a depth of 15 feet. This is done to simulate a demolition trunk line used underwater when clearing a beach of obstacles prior to the arrival of a beach landing force.

To accomplish the assigned task, each student uses a strand of rope approximately 18-20 inches in length. Instructions are clear and concise. Start on the surface treading water with the instructor. Identify yourself and state which knot you are going to tie. The instructor gives the "OK" to start and the student will respond with the "Thumbs down" signal when he is ready to dive down. The instructor will then respond with the "Thumbs down" signal giving the student permission to dive down and tie his first knot on the "Trunk Line." The student is given the option of tying one knot at a time, returning to the surface for air, and repeating the process for each knot, or tying multiple knots on each descent. As part of the descent, the student is required to sink away using a technique that prevents any splashing or noise. Upon reaching the "Trunk Line" the student begins tying the stated knot. Once complete, the student gives the "OK" signal for the knot to be inspected. Upon completion of the knot inspection, and if it is found to be correct, the "OK" signal is then given back to the student. The student then retrieves his rope and asks permission to ascend to the surface by giving the "Thumbs Up" signal to the instructor. After receiving the "Thumbs Up" signal from the instructor the student uses the correct technique for ascending to the surface. This process is repeated over and over until all five knots have been completed. Failure to complete any of the required tasks results in mission failure.

"This is simple, right?" Well… that all depends on your preparation. How much time did you and your swim buddy spend on knot tying and breath holding? Did you work on extending your time each time you attempted a breath hold? How about the stress factors? Did you attempt any breath holds with an elevated heart rate? Did you attempt to tie all five knots in a single breath hold…with your eyes closed? These are just a few of the things that could help in the preparation process for achieving success during this evolution. "What could go wrong?" Well… who knows what your health situation is coming out of Hell week, or better yet… just before the start of the testing process one of your classmates does something stupid and your class is rewarded with 30 minutes of push-ups, sit-ups and 8-count bodybuilders. If you manage to avoid the stupid classmate obstacle, who knows how long you will be treading water before you dive down to take your test. And then there is "The ole leaky dive mask," or worse yet, it just mysteriously gets ripped off of your head in the middle of your knot tying and you are now enjoying the evolution with limited visibility. The one I love the most, because it happened to me, is the missing instructor. You look up to give the "OK" signal to your instructor and he is nowhere to be found. You turn around to look for him and he is floating on his back behind you. You finally get his attention and he takes his sweet time coming over to inspect the knot. He signals you that it is correct and you immediately request permission to go to the surface, but he decides to give you permission in slow motion so that he can extend your breath hold just a few more seconds to see how you handle the situation.

Even with the stress brought on by an elevated heart rate, extended time underwater and limited visibility you can mitigate all of this with a "Slow is smooth, Smooth is fast" mindset due to proper preparation and practice. With the proper preparation and training none of the above mentioned obstacles would make a difference.

In the real world, having a baby is one of those situations that really brings this to light. By doing your research and preparing for the birth of your child, you can alleviate a lot of stress, the type that comes from going through it for the first time and all of the unknown factors, like not knowing the actual time or day.

With the proper training and preparation, you can mitigate the unknown factors that induce stress so when those stressful situations arrive, you can identify them early and immediately go through your procedures or techniques to control the situation or event.

A good way to start that preparation could be taking a childbirth class, like "Lamaze," to educate you and your spouse on what to expect and to learn about the process involved. A good way to deal with the time factor on the actual birth day would be driving the route to the hospital ahead of time to get an idea of how much time it takes to get there, in the morning, during rush hour, and on the weekend. Visit the hospital that you plan to use and familiarize yourself with the correct entrance to use and the easiest route to get to the maternity ward. Also have your "Go bag" (Momma's suitcase) packed and ready to go for the big day.

You should also start preparing the home early for the arrival of the little one. Get all of the items needed for bathing, feeding, and putting the baby down for a nap, all in advance. Ensure all items involved with diaper cleanup are well stocked. Transportation items like infant car seats and strollers need to be researched and acquired ahead of time. Getting familiar with how the car seat and the stroller work prior to using it is a must. You don't want to be learning how to use a lot of these items when you are about to bring the little one home for the first time. This is just the tip of the iceberg of things that can help mitigate potential stress that you may encounter. Educating and preparing yourself can allow you to use the "Slow is smooth, Smooth is fast" concept to identify and mitigate potential stress so you can navigate through it.

The other factor that is sometimes overlooked is the "experience" factor. Experience also plays a major role in the successful use of the "Slow is smooth, Smooth is fast" principle when navigating through those high stress situations. That "Combat Proven" knowledge and experience pays dividends on the confidence side of the house when the poop is hitting the fan. That "Been there, done that" confidence factor is not always a bad thing, as long as you don't let it go to your head. It can be an advantage if used as a proof of concept or a lesson learned experience. I experienced this firsthand as a new guy in the SEAL teams in my first platoon. We were doing Immediate Action Drills (IADs) in the Land Warfare phase of our platoon work up. This type of training tests your ability to react and make decisions when contacted (shot at) by the enemy from various angles, at close range and from afar. The drills are designed to train you on decision making, communications, and maneuvering different types of terrain using various techniques while you engage the enemy. As a new guy, the one thing that always amazed me, was how our Chief always knew what to do and what call to make to maneuver the squad or platoon to give us the best tactical advantage using the terrain and our assets. After awhile, having done multiple runs in various training areas using different types of terrain, it became apparent that the Chief wasn't a genius. He just had more experience, more on the job training than the rest of us.

In the real world, I liken this to doctors, lawyers and accountants. You may be qualified or board certified to practice medicine, law or accounting, but it isn't until you have been exposed to a range of situations in your field or practice, over time, that you truly become "experienced." This "experience" is what allows you to be successful at the "Slow is smooth, Smooth is fast" principle. Experience allows you to anticipate and mitigate potential obstacles and challenges.

Education/Research, *Training* using "Best Practices" coupled with Stress Factors, and *Experience* all play a major role in successfully using a "Slow is smooth, Smooth is fast" mindset.

CHAPTER 6

Plan your Dive, Dive your Plan.

"If you don't design your own life plan, chances are you'll fall into someone else's plan." "And guess what they have planned for you?" "Not much." - *Jim Rohn*

Once you have researched and educated yourself on your overall objective and have committed to achieving it, you are ready for me to introduce you to *Plan your dive, Dive your plan.*

This phrase originates from diving operations, but this theme wasn't chosen because you are breathing underwater. Diving in itself is not that difficult. Trust me on this one. Breathe your gas of choice, know the limits of your diving apparatus and don't hold your breath on ascent. Yeah, I know it's a little more involved than that, but it's not rocket science. The theme of diving is used because of the meticulous planning that is involved with developing a mission where everything hinges on the success of the actual *Dive Plan* from start to finish.

The basic overall concept of the phrase *Plan your dive, Dive your plan* is quite simple. Just "Follow through with your plan!" Yes, it is that simple. The problem is most people don't follow through with any plan. For one reason (excuse) or another they change the plan, avoid the plan when the going gets tough or go off on another tangent all together. The key is to avoid getting fixated on the ever changing latest and greatest trend or taking on *Too much too quickly*.

Let me give you an overview of what is involved with a typical diving operation. The mission begins by inserting the dive pair/s via a supporting asset. Depending upon the environment and/or the enemy situation on the ground this could be by Sea, Air or Land (SEAL). Once at the start point, they submerge to the correct depth and navigate from point to point until they arrive at the objective. This is accomplished by using an "attack board" to navigate on specific bearings for a specific amount of time which has been carefully calculated by each dive pair. This attack board is similar to a kickboard commonly used by swimmers only it doesn't float. It is equipped with a gyro compass, a chronometer and a depth gauge. A successful dive pair flies the bearing as straight as possible while maintaining a good hover to ensure they are on time at each phase of the operation. Bad depth control adds time to each dive leg which in turn can cause problems later on in the *Dive Plan*.

At the objective, there is only a limited amount of time to conduct the mission requirements before safely navigating back to the extract point. Remember, most of the time, these types of operations are conducted at night or in periods of low visibility. This is done to minimize the chance of compromise. If you have never been underwater in the ocean at night, visibility is nearly non-existent. *I have been on some night dives where I could barely see my hand in front of my face, but that's OK, diving operations are not for sightseeing.*

Now that you have somewhat of an overview of what diving operations entail, let's talk about why the theme of diving is used to drive home the concept of following through with the plan.

When developing any type of diving operation, there are a multitude of variables that need to be considered and planned for to ensure mission success. Here are a few that are always at the top of the list.

Time

The number one variable is time. There is only a certain amount of time that is within the hours of limited visibility, so efficient use of it is always the focus.

Light

Any type of operation that requires diving as a key component for mission success is concerned with visibility. These factors include: sunrise, sunset, moonrise and moonset. Surface light either from the sun or moon plays a major role in remaining clandestine throughout any type of operation. How this light affects the divers, at the various intervals throughout the timeline, is always a concern.

Weather

Most of the time, surface weather is not a factor during the actual diving portion of the mission, but a dive always starts with an insert platform somewhere that can be affected by weather.

Every operation that I've been a part of has had a weatherman actively involved with the planning process. This job is not something you normally think of when it comes to planning a Navy SEAL operation, but a good meteorologist is well worth his weight in gold when doing any type of operational planning.

Maritime Factors

These occur daily around the clock when using any type of body of water that is connected to the ocean. Things like tides and currents, if not taken into consideration during the planning process, can guarantee mission failure virtually right out of the gate. Whether the tide is going in or out can tell you which way the current is moving, and depending upon where it is in the process of going in or out, can tell you the speed. Couple that with a good chart study of the operational area and you can glean information that will tell you how that might affect you at various points in the timeline along your underwater route. An experienced dive pair meticulously plans each leg of their route on the surface and underwater so that every maritime variable is taken into consideration when calculating range and bearing, dive time and the gas limits of their dive rigs.

Diving Limitations

Dive physics, dive tables and gas limits have to be followed or diving casualties like Decompression Sickness (DCS) "The Bends," Oxygen Toxicity (excess oxygen in the body tissues which can lead to Central Nervous System problems), Hypercapnia (CO_2 Buildup) or an Arterial Gas Embolism (AGE) (air bubbles in the bloodstream as a result of gross trauma to the lining of the lung due to a rapid ascent while holding your breath) can occur. The majority of diving casualties can be fatal if not attended to/treated immediately (in some cases with a dive chamber). Depending on the length and depth of the dive, the dive pair is going to use the diving apparatus that best fulfills the requirements for that particular mission. Certain diving rigs offer tactical advantages that help prevent the detection of the dive pair. The type and amount of gas a particular diving apparatus uses and holds is also a deciding factor. *"The gas you breathe from your rig doesn't last forever."* The fact that it doesn't last forever is why maritime factors get so much attention in the planning process. A correctly calculated and scheduled mission can take advantage of the benefits provided by certain maritime variables.

Temperature

Depending on the time of year, the temperature of the air and water can play a major role. Every temperature below the normal body temperature range will eventually cause hypothermia over time. There are ways of mitigating temperature, such as wearing wet or dry suits, but even those only prolong the inevitable if the limitations are not adhered to.

Surface Hazards

Boat traffic on the surface can become a concern if the craft has a deep draft (how deep the boat sits in the water). Boat traffic, whether random or scheduled, such as freighters or ferries, need to be researched and planned for in advance so any potential problems can be mitigated.

Marine Life

This is the one variable I always get asked about when the subject of diving at night comes up. The questions are always about sharks, sting rays and sea snakes. Marine life is always a concern and is researched for the area where you are conducting operations. Believe it or not, I've never been bothered or touched by marine life while diving and I've done a lot of diving in shark breeding waters. *One time, on a rare day operation, a tiger shark swam right in front of my buddy and me, but he never paid a bit of attention to us.*

There are a multitude of other variables that can affect the diving operation, but just from the basic ones that I've mentioned you can see how important it is to *Plan your dive and Dive your plan*. Whether it's one dive pair or ten, the plan has to be followed to ensure the safety of the divers and the success of the mission.

Once the dive plan has been developed, contingency planning is then added into the timeline. At each phase of the mission there are certain requirements that have to be met by each dive pair in order to "Charlie Mike" (Continue Mission). Some of these are based off of minimum numbers needed to complete the mission, such as minimum number of personnel needed to conduct the mission objectives, minimum gas pressure in your dive rig at specific phases of the operation and your location along the route at specific times. If any of these are not met, each dive pair has detailed plans and instructions to follow to ensure a safe and secure exfiltration and extraction.

In every operation there is an insertion and infiltration to the target and an exfiltration out of the area to an extraction that takes you back to your base of operations. Not following the "Plan" can cause a *Ripple Effect* across the timeline in each of these phases, starting at the point in time that the change occurred.

If you change your plans en route to the objective, you may get lucky and the change may not affect your ability to reach the target, but let me play the "What if" game with you. "What if" you expended extra gas with your change in plans and you arrived late. You complete your mission requirements at the objective and you start the exfiltration phase, which was originally calculated for the faster current at that point in the timeline. It was planned that way to mitigate the lower gas levels in your rig. Now it's later in the timeline and the current is slower due to the quickly approaching slack tide, which is now going to take you longer to get out of the target area. Then reality sets in as you observe your pressure gauge that now reads near empty. Low readings mean you have to go to the surface, which increases your chance of compromise ten-fold. Upon reaching the surface, you notice the sun is starting to peak over the horizon, which will also aid the enemy in detection. Keep in mind there is a supporting asset standing by somewhere to extract you from the area. *The one that was going to safely deliver you back to your base of operation.* Well, it too has criteria that has to be followed to ensure mission success, and since you are late they are now gone and the back-up plan is in full motion. If I were guessing, that back-up plan calls for a secondary extract at a different location further down the timeline. Do I need to continue? This scenario never gets better.

Plan your dive, Dive your plan and *Execute* the contingency plan when required. If you are going to take the time to develop a solid plan and come up with the contingencies that go with it....then execute as planned. If there are other people involved with the plan such as family, friends, team members, support personnel, co-workers etc. be cognizant of the fact that they are relying on you and everybody involved to follow through with the plan for the best chances of success as a whole.

In the real world, the ripple effect normally doesn't involve fatalities or being captured by the enemy if you change the plan each time you are confronted with an obstacle, but there is a chance the consequences of your actions may affect your family, the organization or the company where you work. Your actions may affect other people's livelihoods and/or the futures of their families. Remember the ripple effect impacts everyone involved from your teammates to support personnel to family members.

I'll be the first to tell you there is no way of planning for absolutely everything. Sometimes things just happen and there's no way you could have planned for or prevented it. Even the best planners in the world sometimes get it wrong. NASA is a prime example. Even with all of their brilliant minds and meticulous planning, there have been major setbacks throughout its history. Things happen, and when they do, go to your contingency plan and adjust accordingly. If you don't have that particular problem in your contingency matrix, use your training and experience to guide you into making the best decision. Keep moving forward, switch to solution mode and figure it out. Success awaits those who are resolute.

Proper preparation is key. Using the rule *Two is one, One is none* can alleviate many of the minor hiccups that come from what I refer to as "The Human Factor." These are the things that get misplaced, damaged or used up early from being left on, etc. If it is something that has a shelf life or duration time, such as batteries, then bring more than one or have an alternate item for the primary one. If you don't *prepare with a spare,* you are only inviting the inevitable.

In the combat environment, a lot of these "spares" are kept in your "Go bag" in case you have to drop everything and evade the situation. In real life, these back up spares are probably in a central location in your house like the pantry, the medicine cabinet or supply drawer. While on the go, these items could be a GPS and Road Atlas, cell phone and charger, jumper cables, a flashlight in the glove box with extra batteries or if you have little ones "The Diaper bag" (The real life "Go bag") which probably includes a lot of these necessities.

In the business world, this starts with big picture items like having a generator for back-up power then moves down from there. In today's e-commerce high-tech world, being up and running at all times is key. Back up personnel, equipment and supplies are so important in this ever-changing landscape.

These are all quick fixes if you plan ahead. Having a standby briefer and laptop ready to go allows for a seamless transition if the portal or video teleconference isn't going as planned. An extra bulb for the Proxima projector or a standalone Wi-Fi hotspot could prevent a delay in the timeline or worse.

Contingency planning is not a new concept, but the key is ensuring that you incorporate it into the overall plan. Schools have been doing this for years to prevent delays in the curriculum learning process by using substitute teachers and prepared lesson plans. Professional sports always have backups for critical positions on the team due to injuries or unforeseen circumstances. Team equipment managers are the masters of being prepared when situations arise to fix, adjust or replace broken equipment at a moment's notice.

Everything that I've covered so far is in reference to planning with "experience." I did this to give you a good overview of what meticulous planning looks like and why you should strive to achieve this level. When you are just beginning the process, you may not have years of experience to look back on to assist you in the development of your initial plan, but you still want to set the standard right from the start. Your first couple of plans for achieving your initial short-term goals don't have to be complex. The plan needs to include contingencies for potential obstacles, but it should be relatively easy to follow as you initially start to move forward.

When you are first starting out, I suggest using the PACE method. PACE stands for Primary, Alternate, Contingency and Emergency. Once you have completed the *Research* and *Educate* phase, you are going to develop a *Primary* plan to meet the objectives required for accomplishing your first set of short term goals.

The *Alternate* portion of this method is more about scheduling than it is about any real changes to the plan. In the real world "Life" happens and you have to adjust accordingly, but all is not lost. Normally it's just a matter of rescheduling that step or event for the next available time. The plan stays the same; the only thing that is adjusted is the timeline. This doesn't mean it's break time or the plan gets put on the back burner. The *Alternate* method is only for adjusting the timeline for the next time slot available after you take care of whatever "Life" handed you.

The *Contingency* method of the PACE plan is where you plan for potential obstacles and what you are going to do if you encounter them. When you first start out, these may be simple things that happen in "Life" or they may be actual events or problems that you can foresee. As you move from short-term goal to short-term goal, your experience level is going to increase and this type of planning, what I sometimes refer to as "What if" planning, will become second nature. As your experience level increases, these contingencies will begin to have criteria that will have to be met before you can continue.

The *Emergency* portion of the PACE method, in the combat environment, is designed to prevent fatalities and save lives. In the real world of "Life," the *Emergency* channel is more of a lifeline to assist you in your movement forward. If you experience a series of setbacks, these are the instructions for what to do next. It is probably going to involve a reset to assess the situation or maybe seek outside help. In the beginning, this is nothing more than shake it off and try again, more practice, extra studying, rehearsals, doing a walk-through or executing a few dry runs. As you move further down the line, you may need to seek outside help such as a professional in that particular field. There is nothing wrong with getting advice or help from a tutor, a mentor, or a consulting expert to assist you in navigating through a difficult area. If you have made realistic attempts at accomplishing a task and you are not having success *Seek Professional Help* to get you over the hurdle. Your initial research should have produced points of contact to reach out to if difficulties arise.

Once you have made the decision to move forward and identified the overall objective, start developing a plan that puts you on course. Allow extra leeway when first starting out to accomplish that first short-term goal. When you begin to move forward, accomplishing that first short term goal is paramount, not just for confidence building, but also for analysis. Accomplishment of each goal allows you to evaluate and assess success. It helps in future planning when it comes to timeline, limitations and contingency planning. It identifies shortfalls and helps in coming up with solutions to mitigate certain obstacles if they present themselves.

Always stayed focused on the plan and keep moving forward. When obstacles arise, adjust your FSF mindset and start developing a solution.

CHAPTER 7

Shoot, Move, and Communicate

"Don't dwell on what went wrong. Instead, focus on what to do next. Spend your energies on moving forward toward finding the answer."
-Denis Waitley

Shoot, Move, and Communicate. These three words are the keys to success in the decision-making process whenever true trouble arises, both on the battlefield and in life. When you are on the ground in a combat environment, this is an all-consuming integral part of your trouble-shooting thought process throughout each and every day. It doesn't matter if you are an infantryman in the big Army or a member of an elite fighting force, this decision matrix works the same. Variables such as priority of the mission, size/type of threat, the size of your element and/or your experience does play a role, but the only true variation from one person or team to the next is how efficiently you execute it.

Whether in life or on the battlefield, the "Shoot" portion of this phrase is the identification process, identifying what the threat or problem is and reacting accordingly. When trouble arises, actions need to be instantaneous; the situation on the ground will determine the call to "Move" towards the threat or away from it. Whatever choice is made, you are moving off the "X" to acquire a better tactical advantage. The "X" is the hot spot, the danger zone, that spot in the middle of the fatal funnel that is not going to get better unless you "Move!"

In a high-threat environment, you are initially moving to take up your field of fire, your area of responsibility, to ensure the group's safety and security. As you do this, you are simultaneously scanning for good cover, a place to get behind, so you can "Shoot," returning effective fire if need be. *Contrary to belief or what you may have seen on television or in the movies, we don't immediately shoot at every gunshot or noise we hear while operating in the red zone.* It is incumbent on everyone in the group to be continuously anticipating the next move. Everyone needs to be looking for that next piece of real estate that offers the best tactical advantage and *C*ommunicating it to the Command and Control (C2) element to assist in the decision-making process.

A good team makes these decisions instantaneously. These actions need to be one controlled fluid movement. Communications need to be calculated and done in a timely manner.

On the battlefield, everything you do, from the time you get out of bed to the time you go back to sleep, is about anticipating the next move, action or response. Whether this is going to chow, standing watch or moving outside the wire, it is a constant for everything you may encounter from point A to point B to point C as you conduct your mission.

Your Course Of Action (COA), what you are going to do in the event of an ambush or attack, encompasses identifying/anticipating the next area or position you can get cover, as you move from point A to point B, so you can return fire if you or your team gets hit here, there or around the corner. This process, this way of thinking, affords a better "Move" decision to be made in high-stress situations. The movement plan is always evolving. It may be to move forward and eliminate the threat, or lay down heavy fire, move away from the threat, reconsolidate the force and move to a better position to assess the situation. If casualties are involved, you may be calling for a Medical Evacuation or a Quick Reaction Force (QRF) to assist you.

Movement also has a lot to do with whether you have a support package. When you have a support package, such as an AC-130 gun ship or set of Apaches that can be called upon to make that threat disappear, then that decision process can sometimes be a little easier. If you are a two-man sniper team with no outside support, you can't afford to be involved in a huge gun battle.

In "Life," this support package can be friends and family, co-workers or outside professionals. The key is keeping them in the loop. Communicating updates or changes in an evolving situation or event can better prepare their response time with any assistance they may be able to provide in helping you with a COA.

A good mission plan has detailed contingency planning criteria that, if followed, will help in the decision-making process to fight or peel out of the situation to fight another way on another day. This isn't the video game world where bullets never run out and you take on anyone and everyone you come in contact with on the battlefield. You only have one life and whether you live to fight another day is dependent upon the choices and decisions you make when problems arise.

The *Shoot, Move, and Communicate* decision-making process is no different in everyday life. Police officers, firefighters and disaster preparedness organizations already have decision-making processes set up that have been developed over time based on past experiences, lessons learned from other agencies and advances in technology. If you are a part of these types of organization, the key is to constantly be training to build your knowledge base and experience level so that when problems arise, this decision making process is one fluid movement and communicated in a timely manner.

For the most part, at some level, most of us already do this throughout the day and don't even realize it. Take driving for an example. When you are looking ahead for obstacles to avoid such as sudden brake lights, people changing lanes without a turn signal, motorists texting and driving, or wildlife, you are already anticipating hazards. Think about this for a minute. How much effort does it take for you to be aware of your surroundings and anticipate potential hazards that you may have to react to while driving? For an experienced driver, this is second nature. It's effortless. Now think back to when you first started driving, most of your concentration was keeping the car between the lines. How many times did someone have to remind you of brake lights ahead or the quickly-approaching stop sign or red light? This effortless anticipation is a direct result of experience behind the wheel. A good experienced driver always has situational awareness so that in the event a hazard, the "X," does present itself, the driver reacts accordingly by slowing down, speeding up or moving away from the "X" and communicating the situation to a higher authority if need be.

In life the initial "Move" could be as simple as changing the conversation due to your anticipation of where it was headed in the first place. Pay attention. Do your homework on the personality, subject or situation you are about to encounter. Anticipate and plan for potential hazards or hiccups. Familiarize yourself or the team with what that potential "X" may look like, so you can train and rehearse to ensure your reactions are instantaneous, smooth and uneventful.

This doesn't just apply to military operations and the business world. You can use this in something as simple as running errands with the kids or going on a trip. By anticipating and identifying potential hazards such as weather, bad traffic areas or construction, you can plan and react accordingly. On long trips, anticipating times for potty breaks and identifying potential eating and lodging areas along the way, ahead of time, can help mitigate these challenges when they present themselves. With kids (and some adults) this process can be as simple as using the "Shiny Object" technique. By anticipating and identifying certain behaviors and responses, you as a parent can sometimes maneuver around a potentially bad situation or a quickly escalating event by changing the attention of the child to something else, the "Shiny Object." You are never going to know everything, but the key to success is always giving yourself the best advantage, and that always begins with proper preparation and planning.

In life, you need to strive to reach a point where proper preparation and planning is the norm for every endeavor you move towards, so that you can mitigate or avoid any potential problem or situation by simultaneously transitioning to a *Shoot, Move, and Communicate* thought process. The end goal is to make your decisions second nature. To be successful, this needs to be an effortless part of your everyday life.

CHAPTER 8

Always Have a Swim Buddy.

"There are two types of people who will tell you that you cannot make a difference in this world: those who are afraid to try and those who are afraid you will succeed." -*Ray Goforth*

Right from day one in BUD/S training, you are informed you will not do anything or go anywhere without a swim buddy. *What is a swim buddy?* A swim buddy is a person who has your back or what we call it in the SEAL teams, "Your Six." A swim buddy is a person who is always looking out for you. On SEAL teams, "swim buddy" is not always synonymous to swimming in the water. It can take on a multitude of meanings. It could mean "shooting buddy" if you are moving together to engage a target. It could mean "dive buddy" if you are navigating underwater on a closed circuit dive rig to do an underwater attack or it could be as simple as a buddy to keep you out of trouble when you are out on the town in a foreign country. Swim buddies always take care of each other and make sure everyone gets back safe and sound.

When you're picking a swim buddy, he or she has to be all in, and sometimes your swim buddy may not be your best friend. In the real world, a swim buddy is a person who is going to help propel you to the next level. These are people who legitimately want you to succeed. They see the future and want to be part of something special. They are passionate about the possibilities, the building process and the terminal objective.

Most of the time, your swim buddy will present himself because he has similar dreams, creative ideas, ambitions or goals. There are no halfway in the water swim buddies. A true swim buddy believes in the plan, is supportive and is constantly providing that motivation to help keep you moving forward to the next short-term goal. Your swim buddy always has your back.

If you are married, you are probably thinking, "my spouse is one of my swim buddies, right?" In a perfect world, I would say yes, but I am realistic and I have no problem calling it the way I see it. Some of you out there have dreams or ideas that involve a terminal objective or overall goal that may not be of interest to your spouse for one reason or another. There are a lot of ideas, ambitions or dreams out there that may have initially seemed impossible before you read this book, but now you know anything is possible with an FSF Mindset. With that in mind, you have to realize it may be hard for a spouse to jump right in those unknown waters with what they may perceive to be your next latest and greatest adventure. That's OK. *I would suggest your spouse read this book,* but if your spouse doesn't meet the swim buddy criteria, all is not lost.

In the real world of "Life," you may not have a swim buddy from day one. Remember your swim buddy is passionate about the plan and all of the possibilities that come with accomplishing the overall objective. On day one, you may not have all of your research done and all of your short term goals identified. This is a process that only requires you to keep moving forward in order to accomplish the next short-term goal en route to the overall objective. If you are passionate about your journey, you will find your swim buddy as you move forward in the process of achieving your goals, however long or short the timeline happens to be. If you are truly passionate about achieving your overall goals, you will attract a swim buddy. You may even attract more than one, or better yet, you may even attract your spouse.

A good swim buddy can take on a variety of roles. It may be a person who is teaching you how to master a skill en route to achieving the next goal. It may be someone who assists you in the preparation for passing a specific qualification test that is required in order to receive that next promotion which leads to your next step in the process. Your swim buddy may even be the one who introduces you to your next mentor or makes that introduction that provides you the opportunity to pitch an idea that will catapult you to the next level.

The overall point is this. You should not be attempting this alone. Surround yourself with positive people, positive friends. Don't engage in negative activity, conversation or discussions. In this game, the glass is always half full and rising. You should always be shooting to keep the glass filling until it overflows and you need a bigger container.

Front Sight Focus

CHAPTER 9

One Force, One Fight

"None of us is as smart as all of us." -*Ken Blanchard*

"Alone we can do so little; together we can do so much." -*Helen Keller*

"We must all hang together or most assuredly we shall all hang separately." -*Benjamin Franklin*

Earlier in my career, prior to going to combat in support of Operation Iraqi Freedom (OIF), I had occasionally heard the phrase "One Force, One Fight." Normally when I heard this, it was part of a military recruiting or advertising campaign. When I arrived in Baghdad for the first time, this phrase really hit home. It was game time for real and not the kind where you played hard for 60 minutes and shook hands with your opponents at the end. These opponents were playing for keeps and there was no way you were going to fight them alone. It didn't matter whether you were a large conventional Army or Marine Unit, a Special Forces Group, a SEAL Team, or a Special Mission Unit. It was just like in Afghanistan; we were fighting an insurgency, indigenous militias and the remnants of the former regime. If we were going to stay alive and come out of this ahead, we were going to have to fight as a "Force." It was going to involve every American service member and contractor, all of the Government Agencies and every Coalition Force we had on the ground.

"One Force, One Fight" means all hands on deck working together to get the problem solved. It's an urgent situation. It's working together with a group of people who normally aren't seen working together, but for a specific project or event, they come together to make it happen. They make it happen to achieve a goal that benefits everyone. When you are truly fighting with a "One Force, One Fight" mindset, you have to think of something bigger than yourself.

In the real world, it may take more than you and your swim buddies to achieve some overall objectives and accomplish your mission. It may require teaming up with other entities with similar goals. You need to learn to play nice with others in this business. You need to master the art of building rapport, handing out genuine compliments and being a good listener. For some of us alpha male types, this is easier said than done. Believe it or not, there are more important things than you in this world. Sometimes you are going to have to put aside that selfish "my way or the highway" approach and be a team player to get to the next level or achieve that next short-term goal. In today's world, it can be difficult to get that all-out effort, no holds barred, just get the job done, participation from subordinates, employees or co-workers. Employees have to believe it's worth their while, "their effort." There are a lot of people out there who are only going to do enough to get by and still receive that paycheck. There has to be "something in it for me" before a lot of people, these days, will put out that extra effort to benefit the organization or accomplish the organization's goals. Sometimes it's as simple as greeting the customer with a smile or showing empathy towards a customer and his or her problems. In today's society, even that is a bridge too far for some employees.

I will tell you this - it's a different time and a different generation. This isn't the 1950s. That pride in your work thing that was supposed to be embedded in you by your parents isn't the norm or the given anymore. When both parents work, there is a high probability that the household is relying on daycare workers and the public school systems to teach pride in your work and good work ethic……good luck with that. I know your pain, but today's generation, or should I say the last three generations, have a totally different mindset, a different set of rules they live by, and it revolves around "What's in it for me if I bust my butt and go above and beyond what is expected?" These generations need instant gratification and acknowledgment for anything and everything they do. It's a by-product of everyone gets a trophy and there's no need to keep score because everyone is a winner. This is a hard pill to swallow if you are the manager or supervisor, and achieving your overall objective involves maximum output and participation from your subordinates, employees or co-workers.

Now having said that, all is not lost. That was my "Buyers beware" speech for all of those who are venturing out for the first time. Not everyone is a "What's in it for me?" type of person. For those who may not be starting from scratch, but are finding it challenging to get "buy in," from their employees or co-workers, etc, to go that extra mile, you may need to establish an incentive plan to jump start all those involved. When I refer to an incentive plan, I'm talking about a genuine reward plan and we all know a true reward plan involves "value." That's right I said it. If you want max output in today's society, it only comes from two things. Either you or your family's lives are in danger or you are going to get something of value when specified goals are achieved. In today's world "value" refers to money or things that can be sold or traded for money or services. I want to drive this point home because employees, subordinates and co-workers are not stupid. If the things they are striving to achieve don't have more value than overtime, your situation just erodes back to the same unproductive environment. I suggest staying away from pins, badges, special parking spots or the opportunity to have your picture on the wall as the rewards in your incentive program. Make your rewards true incentives. There is one word that always gets attention when an incentive program is being introduced: "Bonus." Give them out for performance, at every level, and make sure they are achievable for everyone at each level. If you want to light a fire within your team, your company or your progression en route to achieving your goals, then figure out an incentive that mesmerizes your team or organization.

    Before all of the great business minds of the world start stacking years and years of focus group data outside my front door that suggests otherwise, STOP! I'm not here to demean the "Pins and Badges" program. The majority of the workforce is just fine with it. They are content with their job, their healthcare package, their 401K, etc. and that's okay. There is nothing wrong with that, but that's not the incentive fuel needed to light the fire that I'm talking about. You need to be able to light a spark at every level within the organization from the ground up and you do this with true incentives. I'm talking about the kind that causes an employee, team member or co-worker to think about solutions and ideas to achieving the goals of the organization, 24/7. When the incentive program opens the window to achieving the unimaginable, your organization is going to move forward. Creativity is going to become the company norm. When your incentive program begins in the mailroom and affects every level en route to the CEO, your company has the ability to unite as *"One Force, One Fight."*

CHAPTER 10

Lead In All Situations

"If your actions inspire others to dream more, learn more, do more and become more you are a leader." *–John Quincy Adams*

Lead in all situations. These four words make up the formula that ties the whole process together. The first thing that normally comes to mind when you hear this phrase is "lead by example." If you've had any type of leadership training or held a position of authority, this is not earth shattering news. The idea of "leading by example" has been around since the beginning of time, but when you break this phrase down to its finest point it's much more. To effectively execute the phrase, "lead in all situations," it has to be part of your overall mindset. Everything you do has to revolve around this way of thinking. When your mindset is to "lead in all situations," it's more than just leading by example. This phrase doesn't mean you have to be the leader of the group, but it does require you to lead by example, in all situations, in everything you do, as if the fate of your teammates and the success of the mission depended on it.

When I walked across the quarterdeck on my first SEAL Team, it was a different feeling than I had experienced at previous commands. I was coming aboard to be a member of an elite community, an elite force. I was going to be a part of a group that was created from a plan that had been put in motion by John F

Kennedy and Admiral Arleigh Burke in early 1961, resulting in the commissioning of two SEAL Teams in 1962. What I would soon find out was the SEAL Teams were different; being a member of a SEAL Team came with expectations. Success was expected. Excuses were not going to be tolerated. You would take responsibility for your actions.

At the beginning of the work up cycle for our deployment to Iraq in support of Operation Iraqi Freedom, the Commander addressed all personnel holding key leadership positions. His directions and expectations were clear and concise. He stated that if you wanted to continue to be a member of SEAL Team Two and hold your position of authority, you would "lead in all situations." He also made it clear that it was our job to ensure the troops understood that the expectations were directed at them as well. The implied task was obvious - perform or you would go away. The guidelines were not up for debate, there was no right and left flank, there was no wiggle room. As a US Navy SEAL, I appreciated that type of precise guidance.

Depending upon the mission requirement, your job or position as a US Navy SEAL, can change at a moment's notice. You could be tasked to be a sniper in an over watch position on a helicopter as your teammates took down a suspected ocean freighter, and in the next cycle of darkness, you could be stacked outside a house in downtown Kandahar as an assaulter on an entry team to capture a High Value Target. In both of these situations, you have to be the leader of your position in that operation, no questions asked. It isn't up for debate. Everyone's job on a SEAL team mission from the team leader to the rear security is crucial.

It's no different back in the real world. You have to take pride in everything you do, whether you're the guy in the mailroom or the CEO sitting at the top. It's not only a mindset; it's how you talk and conduct yourself. Leading in all situations starts with the little things. It doesn't matter how small or menial the job. It's about finishing the evolution, the project, or the game. It's about being the best. It's about winning.

There is a saying that is introduced to you during BUD/S training that says it best, "It pays to be a winner." When you and your swim buddy, your boat crew or your squad win any type of competition in BUD/S, you are rewarded with sitting out the next competition or painful evolution because "It pays to be a winner."

A good example is the Goon Squad evolution following the long soft sand runs we would do after our physical training (PT) evolutions. After a long session of what still seems like an unfathomable amount of sit-ups and flutter kicks, push-ups, 8-count body builders, dips, pull-ups and random sprints down to the surf zone to get wet and sandy, (submerging yourself in the ocean then rolling around in the dry sand till you resemble a sugar cookie) we would end the evolution with a long soft sand run in jungle boots. If you haven't experienced running in soft sand with wet clothes and boots, for many miles, I can tell you it ranks up there on the suck factor. Even though the run is unpleasant, that is not the painful part of the evolution. The true test is to avoid being part of the Goon Squad. It is one of many tests during BUD/S training that challenges you both mentally and physically. This evolution has claimed its fair share of quitters.

Here is how it works. Towards the end of the run, the pace instructor begins to pick up speed in an attempt to break out the class. This is done to identify those willing to put out that extra effort and secure a position in the lead pack. As the class starts to spread out and the lead pack has been identified, an instructor establishes the Goon Squad from a point in the line of students stretched down the beach. Once the Goon Squad has been identified, all of the members of the lead pack are told to go sit down, rest, and observe the evolution. "It pays to be a winner!" The remaining members of the class, "The Goon Squad," split up into evenly numbered groups and spread out down the beach at the base of the sand berm that separates the ocean from the compound. The object of each iteration is to win your race, within your group, so you can graduate to the next group that is closest to the compound and repeat the process. Once you win in that final group, you can go join your classmates. Your location when the Goon Squad was identified, determines your group number. If you were in the first 10 or so students after the lead pack, you would be in the first group. Each race normally involves sprints, bear crawling or buddy carries up over the berm and down to the beach, getting wet and sandy and returning to the finish line. The first person back in each group wins. "It pays to be a winner."

This competitive approach plan is forged into you from *Day One, Week One* and is honed throughout your career to keep you sharp. This is done to ensure you are always setting the pace to be part of the lead pack, giving you the best opportunity to win. Competition is an integral part of being a US Navy SEAL. You are expected to perform at your absolute best and secure a position in the lead pack by *Leading In All Situations*.

In today's world, this way of thinking is becoming a lost art. When you hand out ribbons and trophies just for participating, it takes away the drive to succeed. "Why put out and push it to the limit to win the race or put in the extra hours of practice to dominate on the field of play when nobody is going to keep score?" Not wanting to hurt anyone's feelings is not a good reason for this way of thinking. The real world isn't going to give you a trophy unless you earn it. You are not going to get that promotion, that bonus, that anything, unless you put in the time and do the work that produces a product that propels the organization.

Conclusion

"Avoid fragmentation: Find your focus and seek simplicity. Purposeful living calls for elegant efficiency and economy of effort—expending the minimum time and energy necessary to achieve desired goals."
–Dan Millman

Did you notice or get the feeling I was implying that the things I talked about are simple and easy? Occasionally I would actually use the words "Simple" and "Easy" in the text. I did this for a specific reason. I did this to drive home the concept that none of this is "Rocket Science," not even close. The things I talk about are closer to the common sense/simple and easy side of the spectrum, the area where most of us hangout or are located. SEALs aren't successful because they all have an MBA, a PhD or a Law degree. They are successful because they have an FSF mindset and expect success. Their first thought is always to identify the problem, come up with a solution and move forward with a COA. This line of thinking is achievable by anyone and everyone. You don't have to have a Navy SEAL Trident pinned on your chest to acquire this mindset. You just have to believe anything is possible if you come up with a plan and start moving forward, one step at a time. When problems arise take responsibility, no excuses, no whining and never say "I Can't." Rid your mind of negativity and visualize your success. Most success stories don't happen overnight. That's OK. If you are truly passionate about achieving your mission objective you will enjoy the journey.

Don't rush your shot. Ensure your front sight is focused before you pull the trigger to engage each target. Do your homework. *Research* and *Educate* yourself. Don't take on *"Too Much Too Quickly"* as you gain experience to *Anticipate* the next move. Let your training and preparation naturally guide you and work for you so you can make the best decisions as you propel yourself towards mission completion.

Always remember, it's OK, to expect success, to want to be great and to win every time. Regardless what anyone says winning never gets old. I'm here to tell you the majority of the ones at the top and the ones at the bottom want you to stay right where you are currently located. The ones at the top don't want to share the podium and the ones at the bottom don't want to see you get ahead. I know there are exceptions at both ends, but I'm not in the exceptions business. I'm in the reality business and the reality is you can be great and achieve anything if you move forward.

I hope you enjoyed the book. My goal was to pull back the curtain slightly so you could take a peek inside the mind of a warrior and see how that thought matrix works, but even more than that, show you that there are situations in real life where you can take this same thought process and put it to work to help you solve "Life's" everyday problems. It's all about having a plan, executing the plan and when obstacles arise get off the "X." Once you are clear of the "X," *Shoot, Move and Communicate*. It doesn't matter if you're in a firefight or a stressful situation in real life, "Get off of the X." Re-group. Reset. Re-assess and Charlie Mike (Continue Mission). Keep moving forward. Re-acquire the target, that next short-term goal and keep driving towards the finish line.

These two worlds, the "Battlefield" and "Life," can present you with similar situations and a developed FSF mindset can help you *Shoot, Move and Communicate* your way to success.

I want every person who takes that initial step forward to be successful. You can always rest assured that there is at least one person (Me) rooting for you to succeed in achieving your overall mission objective. *Front Sight Focus!*

I will ask you, the reader, for the same request that I recently received from an airline pilot after we had landed early at our destination. He said, "If you liked the flight tell 3 people because I know if you thought it was bad you would tell everyone." With that I ask you for the same request, "If you liked the book please tell 3 people because I know if you thought it was bad you would tell everyone."

David and his team at ETA Solutions LLC are available for speaking engagements. For further information go to www.ETASolutionsLLC.com.

Here is a list of books I recommend to assist you in your pursuit of greatness:

Think and Grow Rich *–Napoleon Hill*

It Starts with Food: Discover the Whole30 and Change Your Life in Unexpected Ways *–Dallas Hartwig and Melissa Hartwig*

How to Make People Like You in 90 Seconds or Less *–Nicholas Boothman*

Fluent Forever: How to Learn Any Language Fast and Never Forget It *–Gabriel Wyner*